HUGE EARTHMOVERS

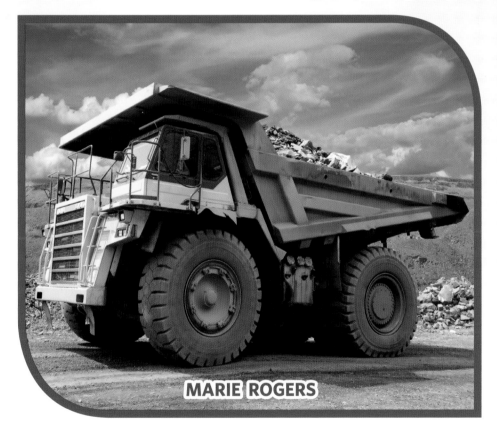

MARIE ROGERS

PowerKiDS
press™

New York

Published in 2022 by The Rosen Publishing Group, Inc.
29 East 21st Street, New York, NY 10010

Portions of this work were originally authored by Kenny Allen and published as *Earthmovers*. All new material in this edition authored by Marie Rogers.

First Edition

Editor: Greg Roza
Cover Design: Michael Flynn
Interior Layout: Rachel Rising

Photo Credits: Cover, pp.1 dragunov/Shutterstock.com; pp. 4, 6, 8, 10, 12, 14, 16, 18, 20, 21 (background) 13Imagery/Shutterstock.com; p. 5 vasiliki/E+/Getty Images; p. 7 ewg3D/E+/Getty Images; p. 9 fotog/Getty Images; p. 11 Avalon_Studio/E+/Getty Images; p. 13 kaband/Shutterstock.com; p.15 TFoxFoto/Shutterstock.com; p. 17 Maksim Safaniuk/Shutterstock.com; p. 19 Avalon_Studio/E+/Getty Images; p. 21 picture alliance/Contributor/Getty Images.

Library of Congress Cataloging-in-Publication Data

Names: Rogers, Marie, 1990- author.
Title: Huge earthmovers / Marie Rogers.
Description: New York : PowerKids Press, [2022] | Series: Big jobs, big
   tools! | Includes index.
Identifiers: LCCN 2020021699 | ISBN 9781725326712 (library binding) | ISBN
   9781725326699 (paperback) | ISBN 9781725326705 (6 pack)
Subjects: LCSH: Earthwork–Juvenile literature. | Excavating
   machinery–Juvenile literature.
Classification: LCC TA732 .R64 2022 | DDC 621.8/65–dc23
LC record available at https://lccn.loc.gov/2020021699

Manufactured in the United States of America

Some of the images in this book illustrate individuals who are models. The depictions do not imply actual situations or events.

CPSIA Compliance Information: Batch #CSPK22. For Further Information contact Rosen Publishing, New York, New York at 1-800-237-9932.

Find us on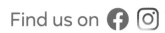

CONTENTS

On the Construction Site

Construction sites are loud and busy! Construction **vehicles** growl and roar. They dig huge holes. They carry heavy loads. They are the earthmovers! Earthmovers are huge machines that do huge jobs.

Big Jobs

Earthmovers do important jobs. Construction workers use them to prepare land for new buildings. Earthmovers are used to build roads. **Mining** companies use earthmovers to dig deep holes and carry dirt and rock away.

Push It!

A bulldozer is a common construction vehicle. It has two tracks that allow it to move on soft dirt. It has a large plate on the front called a **blade**. The blade is used to make land flat for building.

Scoop It!

Wheel loaders look like bulldozers. They have four big wheels. They have a large bucket. Wheel loaders **scoop** up dirt, rock, and snow. The load is often dumped into another vehicle so it can be moved.

Dump It!

Dump trucks are earthmovers with a box on the back. The box tips up to dump its load. Dump trucks **haul** rock, dirt, sand, and pieces of broken buildings. They are also used to carry snow or logs.

Scrape It!

Scrapers are used to make flat ground. The back part is called the hopper. A blade under the hopper scrapes the ground, making it flat. The blade scoops up dirt and rock and puts them in the hopper.

Level It!

Once bulldozers and scrapers do their jobs, a grader is used to make the ground perfectly flat. This is important when making new roads. Graders have a long blade that makes the ground level as it moves forward.

Dig It!

Excavators do many jobs. They have a long, movable arm with a bucket at the end. They're used to dig holes, move heavy things, and knock down old buildings. Some have wheels, and some have tracks.

The Biggest Earthmover

The Bagger 293 is the hugest earthmover in the world! It's a bucket-wheel excavator used to mine coal. It has a giant wheel with buckets on it. The wheel turns, and the buckets scoop up dirt and rock.

Bagger by the Numbers

Largest and heaviest land vehicle: 1,550 tons (1,406 mt)

Height: 310 feet (94.5 m)

Length: 722 feet (220 m)

Size of wheel: 71 feet (21.6 m) wide

Number of buckets on the wheel: 18

Amount each bucket holds: 1,743 gallons (6,600 L)

GLOSSARY

blade: The flat, sharp part of a weapon or tool that is used for cutting.

construction: Having to do with the act of building something.

haul: To pull or carry something heavy.

mine: To dig into the ground to find minerals such as coal, gold, and diamonds.

scoop: To reach under something and lift it.

vehicle: A machine used to carry things from one place to another.

FOR MORE INFORMATION

WEBSITES

Bucket-Wheel Excavator
www.guinnessworldrecords.com/products/books/guinness-world-records-2019/makinghistory/excavator/
Check out this website with some interesting facts about the record-setting Bagger and its Lego replica!

Bulldozer Facts
www.kids.kiddle.co/Bulldozer
Learn more about bulldozers from this informative website.

BOOKS

Arnold, Quinn M. *Bulldozers*. Mankato, MN: Creative Paperbacks, 2018.

Lanier, Wendy Hinote. *Building a Road*. Mendota Heights, MN: Focus Readers, 2020.

INDEX